# Deucalion

# DEUCALION.

COLLECTED STUDIES

OF THE

*LAPSE OF WAVES, AND LIFE OF STONES.*

BY

JOHN RUSKIN, D.C.L., LL.D.,

HONORARY STUDENT OF CHRISTCHURCH, OXFORD; AND HONORARY FELLOW OF CORPUS CHRISTI COLLEGE, OXFORD.

ἐπειὴ μάλα πολλὰ μεταξὺ
οὔρεά τε σκιόεντα, θάλασσά τε ἠχήεσσα·



VOL. II.

NEW YORK:
JOHN WILEY & SONS,
15 ASTOR PLACE.
1886.

# DEUCALION.

## VOL. II.

---

## CHAPTER I.

### LIVING WAVES.

1. THE opening of the second volume of Deucalion with a Lecture on Serpents may seem at first a curiously serpentine mode of advance towards the fulfilment of my promise that the said volume should contain an account of the hills surrounding me at Coniston, (above, vol. i. p. 241, § 38). But I am obliged now in all things to follow in great part the leadings of circumstance: and although it was only the fortuitous hearing of a lecture by Professor Huxley which induced me to take up at present the materials I had by me respecting snake motion, I believe my readers will find their study of undulatory forces dealt through the shattered vertebræ of rocks, very materially enlivened, if not aided, by first observing the transitions of it through the adjusted vertebræ of the serpent. I would rather indeed have made this the matter of a detached essay, but my distinct books are

far too numerous already; and, if I could only complete them to my mind, would in the end rather see all of them fitted into one colubrine chain of consistent strength, than allowed to stand in any broken or diverse relations.

There are, however, no indications in the text of the lecture itself of its possible use in my geological work. It was written as briefly and clearly as I could, for its own immediate purpose: and is given here, as it was delivered, with only the insertion of the passages I was forced to omit for want of time.

2. The lecture, as it stands, was, as I have just said, thrown together out of the materials I had by me; most of them for a considerable time; and with the help of such books as I chanced to possess,—chiefly, the last French edition of Cuvier,—Dr. Russell's Indian Serpents, —and Bell's British Reptiles. Not until after the delivery of the lecture for the second time, was I aware of the splendid work done recently by Dr. Gunther, nor had I ever seen drawings of serpents for a moment comparable, both in action and in detail of scale, to those by Mr. Ford which illustrate Dr. Gunther's descriptions ; or, in colour, and refinement of occasional action, to those given in Dr. Fayrer's Thanatophidia of India. The reader must therefore understand that anything generally said, in the following lecture, of modern scientific shortcoming, or error, is not to be understood as applying to any publication by either of these two authors, who have, I believe, been the first naturalists to adopt the artistically

UNIVERSITY
CALIFORNIA.

Plate VIII.

"Development." Crocodile latent in Toucan.

and mathematically sound method of delineation by plan and profile; and the first to represent serpent action in true lines, whether of actual curve, or induced perspective.

What follows, then, is the text of what I read, or, to the best of my memory, spoke, at the London Institution.

3. In all my lectures on Natural History at Oxford I virtually divided my subject always into three parts, and asked my pupils, first, to consider what had been beautifully thought about the creature; secondly, what was accurately known of it; thirdly, what was to be wisely asked about it.

First, you observe, what was, or had been, beautifully thought about it; the effect of the creature, that is to say, during past ages, on the greatest human minds. *This*, it is especially the business of a gentleman and a scholar to know. It is a king's business, for instance, to know the meaning of the legend of the basilisk, the King of Serpents, who killed with a look, in order that he may not himself become like a basilisk. But that kind of knowledge would be of small use to a viper-catcher.

Then the second part of the animal's history is—what is truly known of it, which one usually finds to be extremely little.

And the third part of its history will be—what remains to be asked about it—what it now behoves us, or will be profitable to us, to discover.

4. It will perhaps be a weight off your minds to be assured that I shall waive to-night the first part of the subject altogether;—except so far as thoughts of it may be suggested to you by Mr. Severn's beautiful introductory diagram,* and by the references I have to make to it, though shown for the sake of the ivy, not the Eve,—its subject being already explained in my Florentine Guide to the Shepherd's Tower. But I will venture to detain you a few moments while I point out how, in one great department of modern science, past traditions may be used to facilitate, where at present they do but encumber, even the materialistic teaching of our own day.

5. When I was furnishing Brantwood, a few years ago, I indulged myself with two bran-new globes, brought up to all the modern fine discoveries. I find, however, that there's so much in them that I can see nothing. The names are too many on the earth, and the stars too crowded in the heaven. And I am going to have made for my Coniston parish school a series of drawings in dark blue, with golden stars, of one constellation at a time, such as my diagram No. 2, with no names written to the stars at all. For if the children don't know their names without print on their diagram, they won't know them without print on the sky. Then there must be a school-manual of the constellations,

---

* The Creation of Eve, bas-relief from the tower of Giotto. The photograph may be obtained from Mr. Ward.

which will have the legend of each told as simply as a fairy tale; and the names of the chief stars given on a map of them, corresponding to the blue diagram,—both of course drawn as the stars are placed in the sky; or as they would be seen on a concave celestial globe, from the centre of it. The having to look down on the stars from outside of them is a difficult position for children to comprehend, and not a very scientific one, even when comprehended.

6. But to do all this rightly, I must have better outlines than those at present extant. The red diagram, No. 3, which has I hope a little amused you, more than frightened, is an enlargement of the outline given on my new celestial globe, to the head of the constellation Draco. I need not tell you that it is as false to nature as it is foolish in art; and I want you to compare it with the uppermost snake head in No. 4, because the two together will show you in a moment what long chapters of 'Modern Painters' were written to explain, —how the real faculty of imagination is always true, and goes straight to its mark: but people with no imagination are always false, and blunder or drivel about their mark. That red head was drawn by a man who didn't know a snake from a sausage, and had no more imagination in him than the chopped pork of which it is made. Of course he didn't know that, and with a scrabble of lines this way and the other, gets together what he thinks an invention—a knot of gratuitous lies, which you con-

tentedly see portrayed as an instrument of your children's daily education. While—two thousand and more years ago—the people who had imagination enough to believe in Gods, saw also faithfully what was to be seen in snakes; and the Greek workman gives, as you see in this enlargement of the silver drachma of Phæstus, with a group of some six or seven sharp incisions, the half-dead and yet dreadful eye, the flat brow, the yawning jaw, and the forked tongue, which are an abstract of the serpent tribe for ever and ever.

And I certify you that all the exhibitions they could see in all London would not teach your children so much of art as a celestial globe in the nursery, designed with the force and the simplicity of a Greek vase.

7. Now, I have done alike with myths and traditions; and perhaps I had better forewarn you, in order, what I am next coming to. For, after my first delivery of this lecture, one of my most attentive hearers, and best accustomed pupils, told me that he had felt it to be painfully unconnected,—with much resultant difficulty to the hearer in following its intention. This is partly inevitable when one endeavours to get over a great deal of ground in an hour; and indeed I have been obliged, as I fastened the leaves together, to cut out sundry sentences of adaptation or transition—and run my bits of train all into one, without buffers. But the actual divisions of what I have to say are clearly jointed for all that; and if you like to jot them down from the leaf I

have put here at my side for my own guidance, these are the heads of them :—

I. Introduction—Imaginary Serpents.
II. The Names of Serpents.
III. The Classification of Serpents.
IV. The Patterns of Serpents.
V. The Motion of Serpents.
VI. The Poison of Serpents.
VII. Caution, concerning their Poison.
VIII. The Wisdom of Serpents.
IX. Caution, concerning their Wisdom.

It is not quite so bad as the sixteenthly, seventeenthly, and to conclude, of the Duke's chaplain, to Major Dalgetty; but you see we have no time to round the corners, and must get through our work as straightly as we may.

We have got done already with our first article, and begin now with the names of serpents; of which those used in the great languages, ancient and modern, are all significant, and therefore instructive, in the highest degree.

8. The first and most important is the Greek 'ophis,' from which you know the whole race are called, by scientific people, ophidia. It means the thing that sees all round; and Milton is thinking of it when he makes the serpent, looking to see if Eve be assailable, say of himself, "Her husband, *for I view far round*, not

near." Satan says that, mind you, in the person of the Serpent, to whose faculties, in its form, he has reduced himself. As an angel, he would have *known* whether Adam was near or not: in the serpent, he has to look and see. This, mind you further, however, is Miltonic fancy, not Mosaic theology;—it is a poet and a scholar who speaks here,—by no means a prophet.

9. Practically, it has never seemed to me that a snake *could* see far round, out of the slit in his eye, which is drawn large for you in my diagram of the rattlesnake;* but either he or the puffadder, I have observed, seem to see with the backs of their heads as well as the fronts, whenever I am drawing them. You will find the question entered into at some length in my sixth lecture in the 'Eagle's Nest'; and I endeavoured to find out some particulars of which I might have given you assurance to-night, in my scientific books; but though I found pages upon pages of description of the scales and wrinkles about snakes' eyes, I could come at no account whatever of the probable range or distinctness in the sight of them; and though extreme pains had been taken to exhibit, in sundry delicate engravings, their lachrymatory glands and ducts, I could neither discover the occasions on which rattlesnakes wept, nor under what consolations they dried their eyes.

* See the careful drawing of the eye of Daboia Russellii, Thanatophidia, p. 14.

10. Next for the word dracon, or dragon. We are accustomed to think of a dragon as a winged and clawed creature; but the real Greek dragon, Cadmus's or Jason's, was simply a serpent, only a serpent of more determined vigilance than the ophis, and guardian therefore of fruit, fountain, or fleece. In that sense of guardianship, not as a protector, but as a sentinel, the name is to be remembered as well fitted for the great Greek lawgiver.

The dragon of Christian legend is more definitely malignant, and no less vigilant. You will find in Mr. Anderson's supplement to my 'St. Mark's Rest,' "The Place of Dragons," a perfect analysis of the translation of classic into Christian tradition in this respect.

11. III. Anguis. The strangling thing, passing into the French 'angoisse' and English 'anguish'; but we have never taken this Latin word for our serpents, because we have none of the strangling or constrictor kind in Europe. It is always used in Latin for the most terrible forms of snake, and has been, with peculiar infelicity, given by scientific people to the most innocent, and especially to those which can't strangle anything. The 'Anguis fragilis' breaks like a tobacco-pipe; but imagine how disconcerting such an accident would be to a constrictor!

12. IV. Coluber, passing into the French 'couleuvre,' a grandly expressive word. The derivation of the Latin one is uncertain, but it will be wise and convenient to reserve it for the expression of coiling. Our word

'coil,' as the French 'cueillir,' is from the Latin 'colligere,' to collect; and we shall presently see that the way in which a snake 'collects' itself is no less characteristic than the way in which it diffuses itself.

13. V. Serpens. The winding thing. This is the great word which expresses the progressive action of a snake, distinguishing it from all other animals; or, so far as modifying the motion of others, making them in that degree serpents also, as the elongated species of fish and lizard. It is the principal object of my lecture this evening to lay before you the law of this action, although the interest attaching to other parts of my subject has tempted me to enlarge on them so as to give them undue prominence.

14. VI. Adder. This Saxon word, the same as nieder or nether, 'the grovelling thing,' was at first general for all serpents, as an epithet of degradation, 'the deaf adder that stoppeth her ears.' Afterwards it became provincial, and has never been accepted as a term of science. In the most scholarly late English it is nearly a synonym with 'viper,' but that word, said to be a contraction for vivipara, bringing forth the young alive, is especially used in the New Testament of the Pharisees, who compass heaven and earth to make one proselyte. The Greek word used in the same place, echidna, is of doubtful origin, but always expresses treachery joined with malice.

15. VII. Snake. German, 'schlange,' the crawling

thing; and with some involved idea of sliminess, as in a snail. Of late it has become partly habitual, in ordinary English, to use it for innocent species of serpents, as opposed to venomous; but it is the strongest and best general term for the entire race; which race, in order to define clearly, I must now enter into some particulars respecting classification, which I find little announced in scientific books.

16. And here I enter on the third division of my lecture, which must be a disproportionately long one, because it involves the statement of matters important in a far wider scope than any others I have to dwell on this evening. For although it is not necessary for any young persons, nor for many old ones, to know, even if they *can* know, anything about the origin or development of species, it is vitally necessary that they should know what a species *is*, and much more what a genus or (a better word) *gens*, a race, of animals is.

17. A gens, race, or kinship, of animals, means, in the truth of it, a group which can do some special thing nobly and well. And there are always varieties of the race which do it in different styles,—an eagle flies in one style, a windhover in another, but both gloriously,—they are 'Gentiles'—gentlemen creatures, well born and bred. So a trout belongs to the true race, or gens, of fish: he can swim perfectly; so can a dolphin, so can a mackerel: they swim in different styles indeed, but they belong-to the true kinship of swimming creatures.

18. Now between the gentes, or races, and between the species, or families, there are invariably links—mongrel creatures, neither one thing nor another,—but clumsy, blundering, hobbling, misshapen things. You are always thankful when you see one that you are not *it*. They are, according to old philosophy, in no process of development up or down, but are necessary, though much pitiable, where they are. Thus between the eagle and the trout, the mongrel or needful link is the penguin. Well, if ever you saw an eagle or a windhover flying, I am sure you must have sometimes wished to be a windhover; and if ever you saw a trout or a dolphin, swimming, I am sure, if it was a hot day, you wished you could be a trout. But did ever anybody wish to be a penguin?

So, again, a swallow is a perfect creature of a true gens; and a field-mouse is a perfect creature of a true gens; and between the two you have an accurate mongrel—the bat. Well, surely some of you have wished, as you saw them glancing and dipping over lake or stream, that you could for half an hour be a swallow: there have been humble times with myself when I could have envied a field-mouse. But did ever anybody wish to be a bat?

19. And don't suppose that you can invert the places of the creatures, and make the gentleman of the penguin, and the mongrel of the windhover,—the gentleman of the bat, and mongrel of the swallow. All these

living forms, and the laws that rule them, are parables, when once you can read; but you can only read them through love, and the sense of beauty; and some day I hope to plead with you a little, of the value of that sense, and the way you have been lately losing it. But as things are, often the best way of explaining the nature of any one creature is to point out the other creatures with whom it is connected, through some intermediate form of degradation. There are almost always two or three, or more, connected gentes, and between each, some peculiar manner of decline and of reascent. Thus, you heard Professor Huxley explain to you that the true snakes were connected with the lizards through helpless snakes, that break like withered branches; and sightless lizards, that have no need for eyes or legs. But there are three other great races of life, with which snakes are connected in other and in yet more marvellous ways. And I do not doubt being able to show you, this afternoon, the four quarters, or, as astrologers would say, the four houses, of the horizon of serpent development, in the modern view, or serpent relation, in the ancient one. In the first quarter, or house, of his nativity, a serpent is, as Professor Huxley showed you, a lizard that has dropped his legs off. But in the second quarter, or house, of his nativity, I shall show you that he is also a duck that has dropped her wings off. In the third quarter, I shall show you that he is a fish that has dropped his fins off. And in the fourth quarter of ascent, or descent, whichever you esteem it, that a serpent is a honeysuckle, with a head put on.

20. The lacertine relations having been explained to you in the preceding lecture by Professor Huxley, I begin this evening with the Duck. I might more easily, and yet more surprisingly, begin with the Dove; but for time-saving must leave your own imaginations to trace the transition, easy as you may think it, from the coo to the quack, and from the walk to the waddle. Yet that is very nearly one-half the journey. The bird is essentially a singing creature, as a serpent is a mute one; the bird is essentially a creature singing for love, as a puffadder is one puffing for anger; and in the descent from the sound which fills that verse of Solomon's Song, "The time of the singing of birds is come, and the voice of the turtle is heard in our land," to the recollection of the last flock of ducks which you saw disturbed in a ditch, expressing their dissatisfaction in that peculiar monosyllable which from its senselessness has become the English expression for foolish talk,* you have actually got down half-way; and in the next flock of geese whom you discompose, might imagine at first you had got the whole way, from the lark's song to the serpent's hiss.

21. But observe, there is a variety of instrumentation

* The substantive 'quack' in its origin means a person who quacks, —i.e., talks senselessly; see Johnson.

in hisses. Most people fancy the goose, the snake, and we ourselves, are alike in the manner of that peculiar expression of opinion. But not at all. Our own hiss, whether the useful and practical ostler's in rubbing down his horse, or that omnipotent one which—please do not try on me just now!—are produced by the pressure of our soft round tongues against our teeth. But neither the goose nor snake can hiss that way, for a goose has got no teeth, to speak of, and a serpent no tongue, to speak of. The sound which imitates so closely our lingual hiss is with them only a vicious and vindictive sigh, —the general disgust which the creature feels at the sight of us expressed in a gasp. Why do you suppose the puffadder is called puffy?* Simply because he swells himself up to hiss, just as Sir Georgius Midas might do to scold his footmen, and then actually and literally 'expires' with rage, sending all the air in his body out at you in a hiss. In a quieter way, the drake and gander do the same thing; and we ourselves do the same thing under nobler conditions, of which presently.

22. But now, here's the first thing, it seems to me, we've got to ask of the scientific people, what use a serpent has for his tongue, since it neither wants it to talk with, to taste with, to hiss with, nor, so far as I know, to lick with,† and least of all to sting with,—and yet, for

* In more graceful Indian metaphor, the 'Father of Tumefaction.' —(Note from a friend.)

† I will not take on me to contradict, but I don't in the least be-

people who do not know the creature, the little vibrating forked thread, flashed out of its mouth, and back again, as quick as lightning, is the most threatening part of the beast; but what is the use of it? Nearly every other creature but a snake can do all sorts of mischief with its tongue. A woman worries with it, a chameleon catches flies with it, a snail files away fruit with it, a humming-bird steals honey with it, a cat steals milk with it, a pholas digs holes in rocks with it, and a gnat digs holes in *us* with it; but the poor snake cannot do any manner of harm with it whatsoever; and what is *his* tongue forked for?

23. I must leave you to find out that at your leisure; and to enter at your pleasure into the relative anatomical questions respecting forms of palate, larynx, and lung, in the dove, the swan, the goose, and the adder,—not unaccompanied by serpentive extension and action in the necks of the hissing birds, which show you what, so to speak, Nature is thinking of. These mechanical questions are all—leather and prunella, or leather and catgut;—the *moral* descent of the temper and meaning in the sound, from a murmur of affection to a gasp of fury, is the real transition of the creature's being. You will find in Kinglake's account of the charge of the Grays in the battle of Balaclava, accurate record of the human

---

lieve, any of the statements about serpents licking their prey before they swallow it.

murmur of long-restrained rage, at last let loose; and may reflect, also at your leisure, on the modes of political development which change a kindly Scot into a fiery dragon.

24. So far of the fall of the bird-angels from song to hiss: next consider for a minute or two the second phase of catastrophe—from walk to waddle. Walk,—or, in prettier creatures still, the run. Think what a descent it is, from the pace of the lapwing, like a pretty lady's, —"Look, where Beatrice, like a lapwing, runs;" or of the cream-coloured courser* of the African desert, whom you might yourselves see run, on your own downs, like a little racehorse, if you didn't shoot it the moment it alighted there,—to the respectable, but, to say the least, unimpressive, gait from which we have coined the useful word to 'waddle.' Can you remember exactly how a duck does walk? You can best fancy it by conceiving the body of a large barrel carried forward on two short legs, and rolling alternately to each side at every step. Once watch this method of motion attentively, and you will soon feel how near you are to dispensing with legs altogether, and getting the barrel to roll along by itself in a succession of zigzags.

25. Now, put the duck well under water, and he *does* dispense with his legs altogether.

There is a bird who—my good friend, and boat-builder,

---

* Cursorius isabellinus (Meyer), Gallicus (Gould).

Mr. Bell, tells me—once lived on Coniston Water, and sometimes visits it yet, called the saw-bill duck, who is the link, on the ducky side; between the ducks and divers: his shape on the whole is a duck's, but his habits are a diver's,—that is to say, he lives on fish, and he catches them deep under water—swimming, under the surface, a hundred yards at a time.

26. We do not at all enough dwell upon this faculty in aquatic birds. Their feet are only for rowing—not for diving. Those little membranous paddles are no use whatever, once under water. The bird's full strength must be used in diving: he dives with his wings—literally flies under water with his wings;—the great northern diver, at a pace which a well-manned boat can't keep up with. The stroke for progress, observe, is the same as in the air; only, in flying under water, the bird has to keep himself down, instead of keeping himself up, and strikes up with the wing instead of down. Well, the great divers hawk at fish this way, and become themselves fish, or saurians, the wings acting for the time as true fins, or paddles. And at the same time, observe, the head takes the shape, and receives the weapons, of the fish-eating lizard.

Magnified in the diagram to the same scale, this head of the sawbill duck (No. 5) is no less terrible than that of the gavial, or fish-eating crocodile of the Ganges. The gavial passes, by the mere widening of the bones of his beak, into the true crocodile,—the crocodile into the serpentine lizard. I drop my duck's wings off through the penguin; and its beak being now a saurian's, I have only to ask Professor Huxley to get rid of its feet for me, and my line of descent is unbroken, from the dove to the cobra, except at the one point of the gift of poison.

27. An important point, you say? Yes; but one which the anatomists take small note of. Legs, or no legs, are by no means the chief criterion of lizard from snake. Poison, or no poison, is a far more serious one. Why should the mere fact of being quadruped, make the creature chemically innocent? Yet no lizard has ever been recognized as venomous.

28. A less trenchant, yet equally singular, law of distinction is found in the next line of relationship we have to learn, that of serpents with fish.

The first quite sweeping division of the whole serpent race is into water serpents and land serpents.* A large number, indeed, like damp places; and I suppose all serpents who ever saw water can swim; but still fix in

---

* Dr. Gunther's division of serpents, ('Reptiles of British India,' p. 166,) the most rational I ever saw in a scientific book, is into five main kinds: burrowing snakes, ground snakes, and tree snakes, on the land; and fresh-water snakes and sea snakes, in the water.

All the water snakes are viviparous; and I believe all the salt-water ones venomous. Of the fresh-water snakes, Dr. Gunther strongly says, "none are venomous," to my much surprise; for I have an ugly recollection of the black river viper in the Zoological Gardens, and am nearly certain that Humboldt speaks of some of the water serpents of Brazil as dangerous.

your minds the intense and broad distinction between the sand asp, which is so fond of heat that if you light a real fire near him he will instantly wriggle up to it and burn himself to death in the ashes, and the water hydra, who lives in the open, often in the deep sea, and though just as venomous as the little fiery wretch, has the body flattened vertically at the tail so as to swim exactly as eels do.

29. Not that I am quite sure that even those who go oftenest to Eel Pie Island quite know how eels *do* swim, and still less how they walk; nor, though I have myself seen them doing it, can I tell you how they manage it. Nothing in animal instinct or movement is more curious than the way young eels get up beside the waterfalls of the Highland streams. They get first into the jets of foam at the edge, to be thrown ashore by them, and then wriggle up the smooth rocks—heaven knows how. If you like, any of you, to put on greased sacks, with your arms tied down inside, and your feet tied together, and then try to wriggle up after them on rocks as smooth as glass, I think even the skilfullest members of the Alpine Club will agree with me as to the difficulty of the feat; and though I have watched them at it for hours, I do not know how much of serpent, and how much of fish, is mingled in the motion. But observe, at all events, there is no walking here on the plates of the belly: whatever motion is got at all, is by undulation of body and lash of tail: so far as by undulation of body, serpentine; so far as by lash of tail, fishy.

30. But the serpent is in a more intimate sense still, a fish that has dropped its fins off. All fish poison is in the fins or tail, not in the mouth. There are no venomous sharks, no fanged pikes; but one of the loveliest fishes of the south coast, and daintiest too when boiled, is so venomous in the fin, that when I was going eagerly to take the first up that came on the fishing-boat's deck with the mackerel line, in my first day of mackerel fishing, the French pilot who was with me caught hold of my arm as eagerly as if I had been going to lay hold of a viper.

Of the common medusa, and of the sting ray, you know probably more than I do: but have any of us enough considered this curious fact; (have any of you seen it stated clearly in any book of natural history?) that throughout the whole fish race,—which, broadly speaking, pass the whole of their existence in one continual gobble,—you never find any poison put into the teeth; and throughout the whole serpent race, never any poison put into the horns, tail, scales, or skin?

31. Besides this, I believe the aquatic poisons are for the most part black; serpent poison invariably white; and, finally, that fish poison is only like that of bees or nettles, numbing and irritating, but not deadly; but that the moment the fish passes into the hydra, and the poison comes through the teeth, the bite is mortal. In these senses, and in many others, (which I could only trace by showing you the undulatory motion of fins in the sea-

horse, and of body in the sole,) the serpent is a fish without fins.

32. Now, thirdly, I said that a serpent was a honeysuckle with a head *put on*. You perhaps thought I was jesting; but nothing is more mysterious in the compass of creation than the relation of flowers to the serpent-tribe,—not only in those to which, in 'Proserpina,' I have given the name Draconidæ, and in which there is recognized resemblance in their popular name, Snapdragon, (as also in the speckling of the Snake's-head Fritillary,) but much more in those carnivorous, insect-eating, and monstrous, insect-begotten, structures, to which your attention may perhaps have been recently directed by the clever caricature of the possible effects of electric light, which appeared lately in the 'Daily Telegraph.' But, seven hundred years ago, to the Florentine, and three thousand years ago, to the Egyptian and the Greek, the mystery of that bond was told in the dedication of the ivy to Dionysus, and of the dragon to Triptolemus. Giotto, in the lovely design which is to-night the only relief to your eyes, thought the story of temptation enough symbolized by the spray of ivy round the hazel trunk; and I have substituted, in my definition, the honeysuckle for the ivy, because, in the most accurate sense, the honeysuckle is an 'anguis'—a strangling thing. The ivy stem increases with age, without compressing the tree trunk, any more than the rock, that it adorns; but the woodbine retains, to a degree not yet measured, but almost, I believe, after

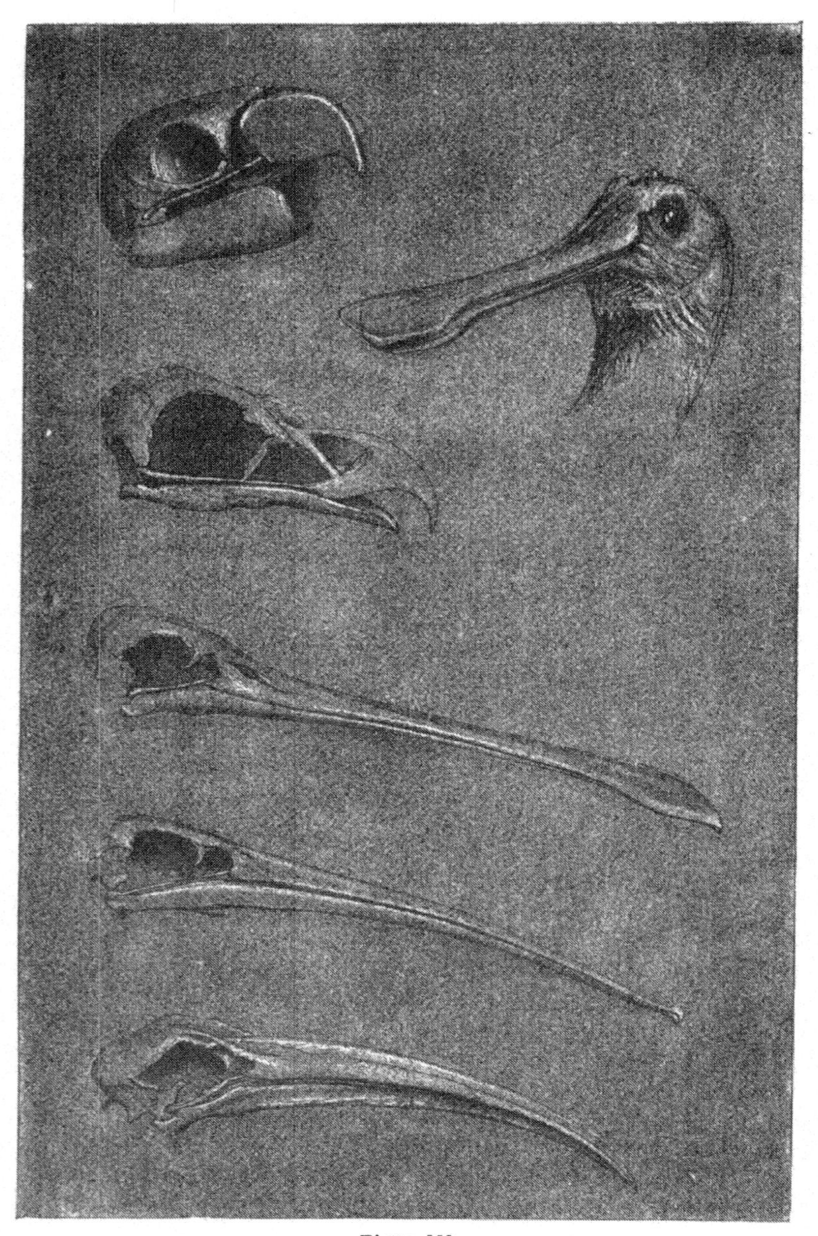

Plate IX.

"Development." Short Noses into Long.

regiment or a centipede may possess, neither body of them can move faster than an individual pair of legs can, —their hundred or thousand feet being each capable of only one step at a time; and, with that allowance, only a certain proportion of pace is possible, and the utmost rapidity of the most active spider, or centipede, does not for an instant equal the dash of a snake in full power. But you—nearly all of you, I fancy—have learned, during the sharp frosts of the last winters, the real secret of it, and will recognize in a moment what the motion is, and only can be, when I show you the real rate of it. It is not often that you can see a snake in a hurry, for he generally withdraws subtly and quietly, even when distinctly seen; but if you put him to his pace either by fear or anger, you will find it is the sweep of the outside edge in skating, carried along the whole body,—that is to say, three or four times over. Outside or inside edge does not, however, I suppose, matter to the snake, the fulcrum being according to the lie of the ground, on the concave or convex side of the curve, and the whole strength of the body is alive in the alternate curves of it.

37. This splendid action, however, you must observe, can hardly ever be seen when the snake is in confinement. Half a second would take him twice the length of his cage; and the sluggish movement which you see there, is scarcely ever more than the muscular extension of himself out of his 'collected' coil into a more or less straight line; which is an action imitable at once with a

coil of rope. You see that one-half of it can move anywhere without stirring the other; and accordingly you may see a foot or two of a large snake's body moving one way, and another foot or two moving the other way, and a bit between not moving at all; which I, altogether, think we may specifically call 'Parliamentary' motion; but this has nothing in common with the gliding and truly serpentine power of the animal when it exerts itself.

38. (Thus far, I stated the matter in my lecture, apologizing at the same time for the incompleteness of demonstration which, to be convincing, would have taken me the full hour of granted attention, and perhaps with small entertainment to most of my hearers. But, for once, I care somewhat to establish my own claim to have first described serpent motion, just as I have cared much to establish Forbes's claim to have first discerned the laws of glacier flow; and I allow myself, therefore, here, a few added words of clearer definition.

39. When languidly moving in its cage, (or stealthily when at liberty,) a serpent may continually be seen to hitch or catch one part of its body by the edge of the scales against the ground, and from the fulcrum of that fixed piece extend other parts or coils in various directions. But this is not the movement of progress. When a serpent is once in full pace, every part of its body moves with equal velocity; and the whole in a series of waves, varied only in sweep in proportion to

not thicker than a tobacco-pipe,—and, according to the most definite account, does not move like ordinary serpents, but throws itself forward a foot or two on the ground, in successive springs, falling in the shape of a horse-shoe. In the five instances given of its bite, death follows, in a boy, ten minutes after the bite; and in the case of two soldiers, bitten by the same snake, but one a minute after the other, in their guard-room, about one in the morning,—the first died at seven in the morning, the second at noon; in both, the powers of sight gradually failing, and they became entirely blind before death. The snake is described as of a dark straw colour, with two black lines behind the head; small, flat head, with *eyes that shone like diamonds.*

43. Next in fatal power to this serpent,—fortunately so rare that I can find no published drawing of it,—come the Cobra, Rattlesnake, and Trigonocephalus, or triangle-headed serpent of the West Indies. Of the last of these snakes, you will find a most terrific account (which I do not myself above one-third believe) in the ninth volume of the English translation of Cuvier's 'Animal Kingdom.' It is a grand book of fifteen volumes, copiously illustrated, and quite unequalled for collection of the things you do not want to know in the body of the text, and for ceasing to be trustworthy the moment it is entertaining. I will read from it a single paragraph concerning the Trigonocephalus, of which you may believe as much or as little as you like. "These reptiles possess an

activity and vivacity of motion truly alarming, A ferocious instinct induces them to dart impetuously upon passengers, either by suddenly letting go the sort of spring which their body forms, rolled in concentric and superpoised circles, and thus shooting like an arrow from the bow of a vigorous archer, or pursuing them by a series of rapid and multiplied leaps, or climbing up trees after them, or even threatening them in a vertical position."

44. The two other serpents, one used to be able to study at our own Zoological Gardens; but the cobra has now for some years had the glass in front of him whitened, to prevent vulgar visitors from poking sticks at him, and wearing out his constitution in bad temper. I do not know anything more disgraceful to the upper classes of England as a body, than that, while on the one hand their chief recreations, without which existence would not be endurable to them, are gambling in horses, and shooting at birds, they are so totally without interest in the naturés and habits of animals in general, that they have never thought of enclosing for themselves a park and space of various kinds of ground, in free and healthy air, in which there should be a perfect gallery, Louvre, or Uffizii, not of pictures, as at Paris, nor of statues, as at Florence, but of living creatures of all kinds, beautifully kept, and of which the contemplation should be granted only to well-educated and gentle people who would take the trouble to travel so far, and

might be trusted to behave decently and kindly to any living creatures, wild or tame.

45. Under existing circumstances, however, the Zoological Gardens are still a place of extreme interest; and I have been able at different times to make memoranda of the ways of snakes there, which have been here enlarged for you by my friends, or by myself; and having been made always with reference to gesture or expression, show you, I believe, more of the living action than you will usually find in scientific drawings: the point which you have chiefly to recollect about the cobra being this curious one—that while the puffadder, and most other snakes, or snakelike creatures, swell when they are angry, the cobra flattens himself; and becomes, for four or five inches of his length, rather a hollow shell than a snake. The beautiful drawing made by Mr. Macdonald in enlarging my sketch from life shows you the gesture accurately, and especially the levelling of the head which gives it the chief terror. It is always represented with absolute truth in Egyptian painting and sculpture; one of the notablest facts to my mind in the entire history of the human race being the adoption by the Egyptians of this serpent for the type of their tyrannous monarchy, just as the cross or the lily was adopted for the general symbol of kinghood by the monarchs of Christendom.

46. I would fain enlarge upon this point, but time forbids me: only please recollect this one vital fact,

that the nature of Egyptian monarchy, however great its justice, is always that of government by cruel force; and that the nature of Christian monarchy is embodied in the cross or lily, which signify either an authority received by divine appointment, and maintained by personal suffering and sacrifice; or else a dominion consisting in recognized gentleness and beauty of character, loved long before it is obeyed.

47. And again, whatever may be the doubtful meanings of the legends invented among all those nations of the earth who have ever seen a serpent alive, one thing is certain, that they all have felt it to represent to them, in a way quite inevitably instructive, the state of an entirely degraded and malignant human life. I have no time to enter on any analysis of the causes of expression in animals, but this is a constant law for them, that they are delightful or dreadful to us exactly in the degree in which they resemble the contours of the human countenance given to it by virtue and vice; and this head of the cerastes, and that of the rattlesnake, are in reality more terrific to you than the others, not because they are more snaky, but because they are more human,—because the one has in it the ghastliest expression of malignant avarice, and the other of malignant pride. In the deepest and most literal sense, to those who allow the temptations of our natural passions their full sway, the curse, fabulously (if you will) spoken on the serpent is fatally and to the full accomplished upon

ourselves; and as for noble and righteous persons and nations, the words are for ever true, "Thou art fairer than the children *of men:* full of grace are thy lips;" so for the ignoble and iniquitous, the saying is for ever true, "Thou art fouler than the children of the Dust, and the poison of asps is under thy lips."

48. Let me show you, in one constant manner of our national iniquity, how literally that is true. Literally, observe. In any good book, but especially in the Bible, you must always look for the literal meaning of everything first,—and act out that, then the spiritual meaning easily and securely follows. Now in the great Song of Moses, in which he foretells, before his death, the corruption of Israel, he says of the wicked race into which the Holy People are to change, "Their wine is the poison of dragons, and the cruel venom of asps." Their wine,—that is to say, of course, not the wine they drink, but the wine they give to drink. So that, as our best duty to our neighbour is figured by the Samaritan who heals wounds by pouring in oil and wine, our worst sin against our neighbour is in envenoming his wounds by pouring in gall and poison. The cruel venom of *Asps*—of that brown gentleman you see there!

49. Now I am sure you would all be very much shocked, and think it extremely wrong, if you saw anybody deliberately poisoning so much as one person in that manner. Suppose even in the interests of science,

to which you are all so devoted, I were myself to bring into this lecture-room a country lout of the stupidest,—the sort whom you produce by Church of England education, and then do all you can to get emigrated out of your way; fellows whose life is of no use to them, nor anybody else; and that—always in the interests of science—I were to lance just the least drop out of that beast's tooth into his throat, and let you see him swell, and choke, and get blue and blind, and gasp himself away—you wouldn't all sit quiet there, and have it so done—would you?—in the interests of science.

50. Well; but how then if in your own interests? Suppose the poor lout had his week's wages in his pocket—thirty shillings or so; and, after his inoculation, I were to pick his pocket of them; and then order in a few more louts, and lance their throats likewise, and pick their pockets likewise, and divide the proceeds of, say, a dozen of poisoned louts, among you all, after lecture: for the seven or eight hundred of you, I could perhaps get sixpence each out of a dozen of poisoned louts; yet you would still feel the proceedings painful to your feelings, and wouldn't take the sixpen'north—would you?

51. But how, if you constituted yourself into a co-operative Egyptian Asp and Mississippi Rattlesnake Company, with an eloquent member of Parliament for the rattle at its tail? and if, brown asps getting scarce, you brewed your own venom of beautiful aspic brown, with a white head, and persuaded your louts to turn their

own pockets inside-out to get it, giving you each sixpence a night,—seven pounds ten a year of lovely dividend!—How does the operation begin to look now? Commercial and amiable—does it not?

52. But how—to come to actual fact and climax—if, instead of a Company, you were constituted into a College of reverend and scholarly persons, each appointed—like the King of Salem—to bring forth the bread and wine of healing knowledge; but that, instead of bread gratis, you gave stones for pay; and instead of wine gratis, you gave asp-poison for pay,—how then? Suppose, for closer instance, that you became a College called of the Body of Christ, and with a symbolic pelican for its crest, but that this charitable pelican had begun to peck—not itself, but other people,—and become a vampire pelican, sucking blood instead of shedding,—how then? They say it's an ill bird that fouls its own nest. My own feeling is that a well-behaved bird will neither foul its own nest nor another's, but that, finding it in any wise foul, it will openly say so, and clean it.

53. Well, I know a village, some few miles from Oxford, numbering of inhabitants some four hundred louts, in which my own College of the Body of Christ keeps the public-house, and therein sells—by its deputy—such poisoned beer that the Rector's wife told me, only the day before yesterday, that she sent for some to take out a stain in a dress with, and couldn't touch the dress with

it, it was so filthy with salt and acid, to provoke thirst; and that while the public-house was there she had no hope of doing any good to the men, who always prepared for Sunday by a fight on Saturday night. And that my own very good friend the Bursar, and we the Fellows, of Corpus, being appealed to again and again to shut up that tavern, the answer is always, "The College can't afford it: we can't give up that fifty pounds a year out of those peasant sots' pockets, and yet 'as a College' live."

Drive that nail home with your own hammers, for I've no more time; and consider the significance of the fact, that the gentlemen of England can't afford to keep up a college for their own sons but by selling death of body and soul to their own peasantry.

54. I come now to my last head of lecture—my caution concerning the wisdom which we buy at such a price. I had not intended any part of my talk to-night to be so grave; and was forced into saying what I have now said by the appointment of Fors that the said village Rector's wife should come up to town to nurse her brother, Mr. Severn, who drew your diagrams for you. I had meant to be as cheerful as I could; and chose the original title of my lecture, 'A Caution to Snakes,' partly in play and partly in affectionate remembrance of the scene in 'New Men and Old Acres,' in which the phrase became at once so startling and so charming, on the lips of my much-regarded friend, Mrs. Kendall.

But this one little bit of caution more I always intended to give, and to give earnestly.

55. What the best wisdom of the Serpent may be, I assume that you all possess;—and my caution is to be addressed to you in that brightly serpentine perfection. In all other respects as wise, in one respect let me beg you to be wiser than the Serpent, and not to eat your meat without tasting it,—meat of any sort, but above all the serpent-recommended meat of knowledge. Think what a delicate and delightful meat that used to be in old days, when it was not quite so common as it is now, and when young people—the best sort of them—really hungered and thirsted for it. *Then* a youth went up to Cambridge, or Padua, or Bonn, as to a feast of fat things, of wines on the lees, well-refined. But now, he goes only to swallow,—and, more's the pity, not even to swallow as a glutton does, with enjoyment; not even—forgive me the old Aristotelian Greek, ἡδομενος τῃ αφῃ—pleased with the going down, but in the saddest and exactest way, as a constrictor does, tasting nothing all the time. You remember what Professor Huxley told you—most interesting it was, and new to me—of the way the great boa does not in any true sense swallow, but only hitches himself on to his meat like a coal-sack;—well, that's the exact way you expect your poor modern student to hitch himself on to *his* meat, catching and notching his teeth into it, and dragging the skin of him tight over it,—till at last—you know I told you a little

ago our artists didn't know a snake from a sausage,—but Heaven help us, your University doctors are going on at such a rate that it will be all we can do, soon, to know a *man* from a sausage.

56. Then think again, in old times what a delicious thing a book used to be in a chimney corner, or in the garden, or in the fields, where one used really to read a book, and nibble a nice bit here and there if it was a bride-cakey sort of book, and cut oneself a lovely slice—fat and lean—if it was a round-of-beef sort of book. But what do you do with a book now, be it ever so good? You give it to a reviewer, first to skin it, and then to bone it, and then to chew it, and then to lick it, and then to give it you down your throat like a handful of pilau. And when you've got it, you've no relish for it, after all. And alas! this continually increasing deadness to the pleasures of literature leaves your minds, even in their most conscientious action, sensitive with agony to the sting of vanity, and at the mercy of the meanest temptations held out by the competition of the schools. How often do I receive letters from young men of sense and genius, lamenting the loss of their strength, and waste of their time, but ending always with the same saying, "I *must* take as high a class as I can, in order to please my father." And the fathers love the lads all the time, but yet, in every word they speak to them, prick the poison of the asp into their young blood, and sicken their eyes with blindness to all the true joys, the true aims, and the

true praises of science and literature; neither do they themselves any more conceive what was once the faith of Englishmen; that the only path of honour is that of rectitude, and the only place of honour, the one that you are fit for. Make your children happy in their youth; let distinction come to them, if it will, after well-spent and well-remembered years; but let them now break and eat the bread of Heaven with gladness and singleness of heart, and send portions to them for whom nothing is prepared;—and so Heaven send you its grace—before meat, and after it.

## CHAPTER II.

### REVISION.

1. If the reader will look back to the opening chapter of 'Deucalion,' he will see that the book was intended to be a collection of the notices of phenomena relating to geology which were scattered through my former works, systematized so far as might be possible, by such additional studies as time permitted me.

Hitherto, however, the scattered chapters have contained nothing else than these additional studies, which, so far from systematizing what preceded them, stand now greatly in need of arrangement themselves; and still more of some explanation of the incidental passages referring to matters of higher science than geology, in which I have too often assumed that the reader is acquainted with—and in some degree even prepared to admit—the modes of thought and reasoning which have been followed throughout the general body of my writings.

I have never given myself out for a philosopher; nor spoken of the teaching attempted in connection with any subject of inquiry, as other than that of a village showman's "Look—and you shall see." But, during the last twenty years, so many baseless semblances of philosophy have announced themselves; and the laws of decent thought and rational question have been so far transgressed (even in our universities, where the moral philosophy they once taught is now only remembered as an obscure tradition, and the natural science in which they are proud, presented only as an impious conjecture), that it is forced upon me, as the only means of making what I have said on these subjects permanently useful, to put into clear terms the natural philosophy and natural theology to which my books refer, as accepted by the intellectual leaders of all past time.

2. To this end, I am republishing the second volume of 'Modern Painters,' which, though in affected language, yet with sincere and very deep feeling, expresses the first and foundational law respecting human contemplation of the natural phenomena under whose influence we exist,—that they can only be seen with their properly belonging joy, and interpreted up to the measure of proper human intelligence, when they are accepted as the work, and the gift, of a Living Spirit greater than our own.

3. Similarly, the moral philosophy which underlies all the appeals, and all the accusations, made in the course of my writings on political science, assumes throughout that the principles of Justice and Mercy which are fastened in the hearts of men, are also expressed in entirely consistent terms throughout the higher—(and even the inferior, when undefiled)—forms of all lovely

literature and art; and enforced by the Providence of a Ruling and Judging Spiritual Power, manifest to those who desire its manifestation, and concealed from those who desire its concealment.

4. These two Faiths, in the creating Spirit, as the source of Beauty,—in the governing Spirit, as the founder and maintainer of Moral Law, are, I have said, *assumed* as the basis of all exposition and of all counsel, which have ever been attempted or offered in my books. I have never held it my duty, never ventured to think of it even as a permitted right, to proclaim or explain these faiths, except only by referring to the writings, properly called inspired, in which the good men of all nations and languages had concurrently—though at far distant and different times—declared them. But it has become now for many reasons, besides those above specified, necessary for me to define clearly the meaning of the words I have used—the scope of the laws I have appealed to, and, most of all, the nature of some of the feelings possible under the reception of these creeds, and impossible to those who refuse them.

5. This may, I think, be done with the best brevity and least repetition, by adding to those of my books still unfinished, 'Deucalion,' 'Proserpina,' 'Love's Meinie,' and 'Fors Clavigera,' explanatory references to the pieces of theology or natural philosophy which have already occurred in each, indicating their modes of connection, and the chiefly parallel passages in the books which are

already concluded; among which I may name the 'Eagle's Nest' as already, if read carefully, containing nearly all necessary elements of interpretation for the others.

6. I am glad to begin with 'Deucalion,' for its title already implies, (and is directly explained in its fifth page as implying,) the quite first principle, with me, of historic reading in divinity, that all nations have been taught of God according to their capacity, and may best learn what farther they would know of Him by reverence for the impressions which He has set on the hearts of each, and all.

I said farther in the same place that I thought it well for the student first to learn the "myths of the Betrayal and Redemption" as they were taught to the heathen world; but I did not say what I meant by the 'Betrayal' and 'Redemption' in their universal sense, as represented alike by Christian and heathen legends.

7. The idea of contest between good and evil spirits for the soul and body of man, which forms the principal subject of all the imaginative literature of the world, has hitherto been the only explanation of its moral phenomena tenable by intellects of the highest power. It is no more a certain or sufficient explanation than the theory of gravitation is of the construction of the starry heavens; but it reaches farther towards analysis of the facts known to us than any other. By '*the* Betrayal' in the passage just referred to I meant the supposed victory, in the present age of the world, of the deceiving spiritual power,

which makes the vices of man his leading motives of action, and his follies, its leading methods. By '*the* Redemption' I meant the promised final victory of the creating and true Spirit, in opening the blind eyes, in making the crooked places straight and the rough plain, and restoring the power of His ministering angels, over a world in which there shall be no more tears.

8. The 'myths'—allegorical fables or stories—in which this belief is represented, were, I went on to say in the same place, "incomparably *truer*" than the Darwinian—or, I will add, any other conceivable materialistic theory—because they are the instinctive products of the natural human mind, conscious of certain facts relating to its fate and peace; and as unerring in that instinct as all other living creatures are in the discovery of what is necessary for their life: while the materialistic theories have been from their beginning products, in the words used in the passage I am explaining (page 5), of the '*half* wits of impertinent multitudes.' They are half-witted because never entertained by any person possessing imaginative power,—and impertinent, because they are always announced as if the very defect of imagination constituted a superiority of discernment.

9. In one of the cleverest—(and, in description of the faults and errors of religious persons, usefullest)—books of this modern half-witted school, "une cure du Docteur Pontalais," of which the plot consists in the revelation by an ingenious doctor to an ingenuous priest that the

creation of the world may be sufficiently explained by dropping oil with dexterity out of a pipe into a wine-glass,—the assumption that 'la logique' and 'la methode' were never applied to theological subjects except in the Quartier Latin of Paris in the present blessed state of Parisian intelligence and morals, may be I hope received as expressing nearly the ultimate possibilities of shallow arrogance in these regions of thought; and I name the book as one extremely well worth reading, first as such; and secondly because it puts into the clearest form I have yet met with, the peculiar darkness of materialism, in its denial of the hope of immortality. The hero of it, who is a perfectly virtuous person, and inventor of the most ingenious and benevolent machines, is killed by the cruelties of an usurer and a priest; and in dying, the only consolation he offers his wife and children is that the loss of one life is of no consequence in the progress of humanity.

This unselfish resignation to total death is the most heroic element in the Religion now in materialist circles called the Religion "of Humanity," and announced as if it were a new discovery of nineteenth-century sagacity, and able to replace in the system of its society, alike all former ideas of the power of God, and destinies of man.

10. But, in the first place, it is by no means a new discovery. The fact that the loss of a single life is of no consequence when the lives of many are to be saved, is,

and always has been, the root of every form of beautiful courage; and I have again and again pointed out, in passages scattered through writings carefully limited in assertion, between 1860 and 1870, that the heroic actions on which the material destinies of this world depend are almost invariably done under the conception of death as a calamity, which is to be endured by one for the deliverance of many, and after which there is no personal reward to be looked for, but the gratitude or fame of which the victim anticipates no consciousness.

11. In the second place, this idea of self-sacrifice is no more sufficient for man than it is new to him. It has, indeed, strength enough to maintain his courage under circumstances of sharp and instant trial; but it has no power whatever to satisfy the heart in the ordinary conditions of social affection, or to console the spirit and invigorate the character through years of separation or distress. Still less can it produce the states of intellectual imagination which have hitherto been necessary for the triumphs of constructive art; and it is a distinctive essential point in the modes of examining the arts as part of necessary moral education, which have been constant in my references to them, that those of poetry, music, and painting, which the religious schools who have employed them usually regard only as stimulants or embodiments of faith, have been by me always considered as its *evidences*. Men

do not sing themselves into love or faith; but they are incapable of true song, till they love, and believe.

12. The lower conditions of intellect which are concerned in the pursuit of natural science, or the invention of mechanical structure, are similarly, and no less intimately, dependent for their perfection on the lower feelings of admiration and affection which can be attached to material things: these also—the curiosity and ingenuity of man—live by admiration and by love; but they differ from the imaginative powers in that they are concerned with things seen—not with the evidences of things unseen—and it would be well for them if the understanding of this restriction prevented them in the present day as severely from speculation as it does from devotion.

13. Nevertheless, in the earlier and happier days of Linnæus, de Saussure, von Humboldt, and the multitude of quiet workers on whose secure foundation the fantastic expatiations of modern science depend for whatever good or stability there is in them, natural religion was always a part of natural science; it becomes with Linnæus a part of his definitions; it underlies, in serene modesty, the courage and enthusiasm of the great travellers and discoverers, from Columbus and Hudson to Livingstone; and it has saved the lives, or solaced the deaths, of myriads of men whose nobleness asked for no memorial but in the gradual enlargement of the realm of manhood, in habitation, and in social virtue.

14. And it is perhaps, of all the tests of difference between the majestic science of those days, and the wild theories or foul curiosities of our own, the most strange and the most distinct, that the practical suggestions which are scattered through the writings of the older naturalists tend always directly to the benefit of the general body of mankind; while the discoverers of modern science have, almost without exception, provoked new furies of avarice, and new tyrannies of individual interest; or else have directly contributed to the means of violent and sudden destruction, already incalculably too potent in the hands of the idle and the wicked.

15. It is right and just that the reader should remember, in reviewing the chapters of my own earlier writings on the origin and sculpture of mountain form, that all the investigations undertaken by me at that time were connected in my own mind with the practical hope of arousing the attention of the Swiss and Italian mountain peasantry to an intelligent administration of the natural treasures of their woods and streams. I had fixed my thoughts on these problems where they are put in the most exigent distinctness by the various distress and disease of the inhabitants of the valley of the Rhone, above the lake of Geneva: a district in which the adverse influences of unequal temperatures, unwholesome air, and alternate or correlative drought and inundation, are all gathered in hostility against a

race of peasantry, the Valaisan, by nature virtuous, industrious, and intelligent in no ordinary degree, and by the hereditary and natural adversities of their position, regarded by themselves as inevitable, reduced indeed, many of them, to extreme poverty and woful disease; but never sunk into a vicious or reckless despair.

16. The practical conclusions at which I arrived, in studying the channels and currents of the Rhone, Ticino, and Adige, were stated first in the letters addressed to the English press on the subject of the great inundations at Rome in 1871 ('Arrows of the Chase,' vol. ii., pp. 111–120), and they are again stated incidentally in 'Fors' (Letter XIX., pp. 239, 240), with direct reference to the dangerous power of the Adige above Verona. Had those suggestions been acted upon, even in the most languid and feeble manner, the twentieth part of the sums since spent by the Italian government in carrying French Boulevards round Tuscan cities, and throwing down their ancient streets to find lines for steam tramways, would not only have prevented the recent inundations in North Italy, but rendered their recurrence for ever impossible.

17. As it is thus the seal of rightly directed scientific investigation, to be sanctified by loving anxiety for instant practical use, so also the best sign of its completeness and symmetry is in the frankness of its communication to the general mind of well-educated persons.

The fixed relations of the crystalline planes of minerals, first stated, and in the simplest mathematical terms expressed, by Professor Miller of Cambridge, have been examined by succeeding mineralogists with an ambitious intensity which has at last placed the diagrams of zone circles for quartz and calcite, given in Cloizeaux's mineralogy, both as monuments of research, and masterpieces of engraving, a place among the most remarkable productions of the feverish energies of the nineteenth century. But in the meantime, all the characters of minerals, except the optical and crystalline ones, which it required the best instruments to detect, and the severest industry to register, have been neglected;* the arrangement of collections in museums has been made unintelligibly scientific, without the slightest consideration whether the formally sequent specimens were in lights, or places, where they could be ever visible; the elements of mineralogy prepared for schools have been diversified by eight or ten different modes, nomenclatures, and systems of notation; and while thus the study of mineralogy at all has become impossible to young

* Even the chemistry has been allowed to remain imperfect or doubtful, while the planes of crystals were being counted: thus for an extreme instance, the most important practical fact that the colour of ultramarine is destroyed by acids, will not be found stated in the descriptions of that mineral by either Miller, Cloizeaux, or Dana; and no microscropic studies of refraction have hitherto informed the public why a ruby is red, a sapphire blue, or a flint black. On a large scale, the darkening of the metamorphic limestones, near the central ranges, remains unexplained.

people, except as a very arduous branch of mathematics, that of its connection with the structure of the earth has been postponed by the leading members of the Geological Society, to inquire into the habits of animalculæ fortunately for the world invisible, and monsters fortunately for the world unregenerate. The race of old Swiss guides, who knew the flowers and crystals of their crags, has meanwhile been replaced by chapmen, who destroy the rarest living flowers of the Alps to raise the price of their herbaria, and pedestrian athletes in the pay of foolish youths; the result being that while fifty years ago there was a good and valuable mineral cabinet in every important mountain village, it is impossible now to find even at Geneva anything offered for sale but dyed agates from Oberstein; and the confused refuse of the cheap lapidary's wheel, working for the supply of Mr. Cooke's tourists with 'Trifles from Chamouni.'

18. I have too long hoped to obtain some remedy for these evils by putting the questions about simple things which ought to be answered in elementary schoolbooks of science, clearly before the student. My own books have thus sometimes become little more than notes of interrogation, in their trust that some day or other the compassion of men of science might lead them to pause in their career of discovery, and take up the more generous task of instruction. But so far from this, the compilers of popular treatises have sought always to make them more saleable by bringing them up to the level

of last month's scientific news; seizing also invariably, of such new matter, that which was either in itself most singular, or in its tendencies most contradictory of former suppositions and credences: and I purpose now to redeem, so far as I can, the enigmatical tone of my own books, by collecting the sum of the facts they contain, partly by indices, partly in abstracts, and so leaving what I myself have seen or known, distinctly told, for what use it may plainly serve.

For a first step in the fulfilment of this intention, some explanation of the circumstances under which the preceding lecture (on the serpent) was prepared, and of the reasons for its insertion in 'Deucalion,' are due to the reader, who may have thought it either careless in its apparent jesting, or irrelevant in its position.

I happened to be present at the lecture given on the same subject, a few weeks before, by Professor Huxley, in which the now accepted doctrine of development was partly used in support of the assertion that serpents were lizards which had lost their legs; and partly itself supported reciprocally, by the probability which the lecturer clearly showed to exist, of their being so.

Without denying this probability, or entering at all into the question of the links between the present generation of animal life and that preceding it, my own lecture was intended to exhibit another series, not of merely probable, but of observable, facts, in the relation of living animals to each other.

And in doing so, to define, more intelligibly than is usual among naturalists, the disputed idea of Species itself.

As I wrote down the several points to be insisted on, I found they would not admit of being gravely treated, unless at extreme cost of pains and time—not to say of weariness to my audience. Do what I would with them, the facts themselves were still superficially comic, or at least grotesque: and in the end I had to let them have their own way; so that the lecture accordingly became, apparently, rather a piece of badinage suggested by Professor Huxley's, than a serious complementary statement.

Nothing, however, could have been more seriously intended; and the entire lecture must be understood as a part, and a very important part, of the variously reiterated illustration, though all my writings, of the harmonies and intervals in the being of the existent animal creation—whether it be developed or undeveloped.

The nobly religious passion in which Linnæus writes the prefaces and summaries of the 'Systema Naturæ,' with the universal and serene philanthropy and sagacity of Humboldt, agree in leading them to the optimist conclusion, best, and unsurpassably, expressed for ever in Pope's 'Essay on Man'; and with respect to lower creatures, epigrammatized in the four lines of George Herbert,—

> "God's creatures leap not, but express a feast
> Where all the guests sit close, and nothing wants.
> Frogs marry fish and flesh ;—bats, bird and beast,
> Sponges, non-sense and sense, mines,* th' earth and plants."

* 'Mines' mean crystallized minerals.

And the thoughts and feelings of these, and all other good, wise, and happy men, about the world they live in, are summed in the 104th Psalm.

On the other hand, the thoughts of cruel, proud, envious, and unhappy men, of the Creation, always issue out of, and gather themselves into, the shambles or the charnel house: the word 'shambles,' as I use it, meaning primarily the battle-field, and secondly, every spot where any one rejoices in taking life; * and the 'charnel house' meaning collectively, the Morgue, brothel, and vivisection-room.

But, lastly, between these two classes, of the happy and the heartless, there is a mediate order of men both unhappy and compassionate, who have become aware of another form of existence in the world, and a domain of zoology extremely difficult of vivisection,—the diabolic. These men, of whom Byron, Burns, Goethe, and Carlyle are in modern days the chief, do not at all feel that the Nature they have to deal with expresses a Feast only; or that her mysteries of good and evil are reducible to a quite visible Kosmos, as they stand; but that there is another Kosmos, mostly invisible, yet perhaps tangible, and to be felt if not seen.†

* Compare the Modern with the Ancient Mariner—gun versus cross-bow.—" A magnificent albatross was soaring about at a short distance astern, for some time in the afternoon, and was knocked over, but unfortunately not picked up." ('Natural History of the Strait of Magellan'; Edmonston and Douglas, 1871, page 225.)

† 'The Devil his Origin Greatness and Decadence,' (*Sic*, without commas,) Williams and Norgate, 1871.

Without entering, with Dr. Reville of Rotterdam, upon the question how men of this inferior quality of intellect become possessed either of the idea—or substance—of what they are in the habit of calling 'the Devil'; nor even into the more definite historical question, "how men lived who did seriously believe in the Devil"—(that is to say, every saint and sinner who received a decent education between the first and the seventeenth centuries of the Christian æra,)—I will merely advise my own readers of one fact respecting the above-named writers, of whom, and whose minds, I know somewhat more than Dr. Reville of Rotterdam,—that *they*, at least, do not use the word 'Devil' in any metaphorical, typical, or abstract sense, but—whether they believe or disbelieve in what they say—in a distinctly personal one: and farther, that the conceptions or imaginations of these persons, or any other such persons, greater or less, yet of their species—whether they are a mere condition of diseased brains, or a perception of really existent external forces,—are nevertheless real *Visions*, described by them 'from the life,' as literally and straightforwardly as ever any artist of Rotterdam painted a sot—or his pot of beer: and farther—even were we at once to grant that all these visions—as for instance Zechariah's, "I saw the Lord sitting on His Throne, and Satan standing at His right hand to resist Him," *are* nothing more than emanations of the unphosphated nervous matter—still, these states of delirium are an essential part of human natural history: and the species of human Animal subject to them, with the peculiar characters of the phantoms which result from its diseases of the brain, are a much more curious and important subject of science than that which principally occupies the scientific mind of modern days—the species of vermin which are the product of peculiar diseases of the skin.

I state this, however, merely as a necessary Kosmic principle, without any intention of attempting hencefor-ward to engage my readers in any department of Natural History which is outside of the ordinary range of Optics and Mechanics: but if they should turn back to passages of my earlier books which did so, it must always be un-derstood that I am just as literal and simple in language as any of the writers above referred to: and that, for in-stance, when in the first volume of 'Deucalion,' p. 206, I say of the Mylodon—"This creature the Fiends delight to exhibit to you," I don't mean by 'the Fiends' my good and kind geological friends at the British Museum, nor even the architect who made the drain-pipes from the posteriors of its gargoyles the principal shafts in his design for the front of the new building,—be it far from me,—but I do mean, distinctly, Powers of supernatural Mischief, such as St. Dunstan, or St. Anthony, meant by the same expressions.

With which advice I must for the present end this bit of explanatory chapter, and proceed with some of the glacial investigations relating only to the Lakes—and not to the inhabitants—whether of Coniston or Caina.

## CHAPTER III.

### BRUMA ARTIFEX.

1. The frost of 9th March, 1879, suddenly recurrent and severe, after an almost Arctic winter, found the soil and rock of my little shaded hill garden, at Brantwood, chilled underneath far down; but at the surface, saturated through every cranny and pore with moisture, by masses of recently thawed snow.

The effect of the acutely recurrent frost on the surface of the gravel walks, under these conditions, was the tearing up of their surface as if by minutely and delicately explosive gases; leaving the heavier stones imbedded at the bottom of little pits fluted to their outline, and raising the earth round them in a thin shell or crust, sustained by miniature ranges of basaltic pillars of ice, one range set above another, with level plates or films of earth between; each tier of pillars some half-inch to an inch in height, and the storied architecture of them two or three inches altogether; the little prismatic crystals of which each several tier was composed being sometimes knit into close masses with radiant silky lustre, and sometimes separated into tiny, but innumerable shafts, or needles, none more than the twentieth of an inch thick, and many terminating in needle-*points*, of extreme fineness.

2. The soft mould of the garden beds, and the crumbling earth in the banks of streams, were still more singularly divided. The separate clods,—often the separate *particles*,—were pushed up, or thrust asunder, by thread-like crystals, *contorted* in the most fantastic lines, and presenting every form usual in twisted and netted chalcedonies, except the definitely fluent or meltingly diffused conditions, here of course impossible in crystallizations owing their origin to acute and steady frost. The coils of these minute fibres were also more parallel in their swathes and sheaves than chalcedony; and more lustrous in their crystalline surfaces: those which did not sustain any of the lifted clods, usually terminating in fringes of needle-points, melting beneath the breath before they could be examined under the lens.

3. The extreme singularity of the whole structure lay, to my mind, in the fact that there was nowhere the least vestige of *stellar* crystallization. No resemblance could be traced,—no connection imagined,—between these coiled sheaves, or pillared aisles, and the ordinary shootings of radiant films along the surface of calmly freezing water, or the symmetrical arborescence of hoar-frost and snow. Here was an ice-structure wholly of the earth, earthy; requiring for its development, the weight, and for its stimulus, the interference, of clods or particles of earth. In some places a small quantity of dust, with a

large supply of subterranean moisture, had been enough to provoke the concretion of masses of serpentine filaments three or four inches long; but where there was no dust, there were no filaments, and the ground, whether dry or moist, froze hard under the foot.

4. Greatly blaming myself for never having noticed this structure before, I have since observed it, with other modes of freezing shown in the streamlets of the best watered district of the British Islands,—with continually increasing interest: until nearly all the questions I have so long vainly asked myself and other people, respecting the *variable* formations of crystalline minerals, seem to me visibly answerable by the glittering, and softly by the voice, of even the least-thought-of mountain stream, as it relapses into its wintry quietness.

5. Thus, in the first place, the action of common opaque white quartz in filling veins, caused by settlement or desiccation, with transverse threads, imperfectly or tentatively crystalline, (those traversing the soft slates of the Buet and Col d'Anterne are peculiarly characteristic, owing to the total absence of lustrous surface in the filaments, and the tortuous aggregation of their nearly solidified tiers or ranks,) cannot but receive some new rays of light in aid of its future explanation, by comparison with the agency here put forth, before our eyes, in the early hours of a single frosty morning; agency almost measurable in force and progress, resulting in the steady elevation of pillars of ice, bearing up an earthy

roof, with strength enough entirely to conquer its adherence to heavier stones imbedded in it.

6. Again. While in its first formation, lake or pool ice throws itself always, on calm water, into stellar or plumose films, shot in a few instants over large surfaces; or, in small pools, filling them with spongy reticulation as the water is exhausted, the final structure of its compact mass is an aggregation of vertical prisms, easily separable, when thick ice is slowly thawing: prisms neither formally divided, like those of basalt, nor in any part of their structure founded on the primitive hexagonal crystals of the ice; but starch-like, and irregularly acute-angled.

7. Icicles, and all other such accretions of ice formed by additions at the surface, by flowing or dropping water, are always, when unaffected by irregular changes of temperature or other disturbing accidents, composed of exquisitely transparent vitreous ice, (the water of course being supposed transparent to begin with)—compact, flawless, absolutely smooth at the surface, and presenting on the fracture, to the naked eye, no evidence whatever of crystalline structure. They will enclose living leaves of holly, fern, or ivy, without disturbing one fold or fringe of them, in clear jelly (if one may use the word of anything frozen so hard), like the daintiest candyings by Parisian confectioner's art, over glacé fruit, or like the fixed juice of the white currant in the perfect confiture of Bar-le-Duc;—and the frozen gelatine melts, as it

forms, stealthily, serenely, showing no vestige of its crystalline power; pushing nowhere, pulling nowhere; revealing in dissolution, no secrets of its structure; affecting flexile branches and foliage only by its weight, and letting them rise when it has passed away, as they rise after being bowed under rain.

8. But ice, on the contrary, formed by an unfailing supply of running water over a rock surface, increases, not from above, but *from beneath*. The stream is never displaced by the ice, and forced to run over it, but the ice is always lifted by the stream; and the tiniest runlet of water keeps its own rippling way on the rock as long as the frost leaves it life to run with. In most cases, the tricklings which moisten large rock surfaces are supplied by deep under-drainage which no frost can reach; and then, the constant welling forth and wimpling down of the perennial rivulet, seen here and there under its ice, glittering, in timed pulses, steadily, and with a strength according to the need, and practically infinite, heaves up the accumulated bulk of chalcedony it has formed, in masses a foot or a foot and a half thick, if the frost hold; but always more or less opaque in consequence of the action of the sun and wind, and the superficial additions by adhering snow or sleet; until the slowly nascent, silently uplifted, but otherwise motionless glaciers,—here taking casts of the crags, and fitted into their finest crannies with more than sculptor's care, and anon extended in rugged undulation over moss or shale, cover the oozy

slopes of our moorlands with *statues* of cascades, where, even in the wildest floods of autumn, cascade is not.

9. Actual waterfalls, when their body of water is great, and much of it reduced to finely divided mist, build or block themselves up, during a hard winter, with disappointingly ponderous and inelegant incrustations,—I regret to say more like messes of dropped tallow than any work of water-nymphs. But a small cascade, falling lightly, and shattering itself only into *drops*, will always do beautiful things, and often incomprehensible ones. After some fortnight or so of clear frost in one of our recent hard winters at Coniston, a fall of about twenty-five feet in the stream of Leathes-water, beginning with general glass basket-making out of all the light grasses at its sides, built for itself at last a complete veil or vault of finely interwoven ice, under which it might be seen, when the embroidery was finished, falling tranquilly: its strength being then too far subdued to spoil by overloading or over-labouring the poised traceries of its incandescent canopy.

10. I suppose the component substance of this vault to have been that of ordinary icicle, varied only in direction by infinite accidents of impact in the flying spray. But without including any such equivocal structures, we have already counted five stages of ice familiar to us all, yet not one of which has been accurately described, far less explained. Namely,

(1) Common deep-water surface ice, increased from

beneath, and floating, but, except in the degrees of its own expansion, not uplifted.

(2) Surface ice on pools of streams, *exhausting* the water as it forms, and adherent to the stones at its edge. Variously increased in crusts and films of spongy network.

(3) Ice deposited by external flow or fall of water in superadded layers—exogen ice,—on a small scale, vitreous, and perfectly compact, on a large one, coarsely stalagmitic, like impure carbonate of lime, but I *think* never visibly fibrous-radiant, as stalactitic lime is.

(4) Endogen ice, formed from beneath by tricklings over ground surface.

(5) Capillary ice, extant from pores in the ground itself, and carrying portions of it up with its crystals.

11. If to these five modes of slowly progressive formation, we add the swift and conclusive arrest of vapour or dew on a chilled surface, we shall have, in all, six different kinds of—terrestrial, it may be called as opposed to aerial—congelation of water: exclusive of all the atmospheric phenomena of snow, hail, and the aggregation of frozen or freezing particles of vapour in clouds. Inscrutable these, on our present terms of inquiry; but the six persistent conditions, formed before our eyes, may be examined with some chance of arriving at useful conclusions touching crystallization in general.

12. Of which, this universal principle is to be first understood by young people;—that every crystalline

substance has a brick of a particular form to build with, usually, in some angle or modification of angle, quite the mineral's own special property,—and if not absolutely peculiar to it, at least peculiarly used by it. Thus, though the brick of gold, and that of the ruby-coloured oxide of copper, are alike cubes, yet gold grows trees with its bricks, and ruby copper weaves samite with them. Gold cannot plait samite, nor ruby copper branch into trees; and ruby itself, with a far more convenient and adaptable form of brick, does neither the one nor the other. But ice, which has the same form of bricks to build with as ruby, can, at its pleasure, bind them into branches, or weave them into wool; buttress a polar cliff with adamant, or flush a dome of Alp with light lovelier than the ruby's.

13. You see, I have written above, 'ruby,' as I write 'gold' or ice, not calling their separate crystals, rubies, or golds, or ices. For indeed the laws of structure hitherto ascertained by mineralogists have not shown us any essential difference between substances which crystallize habitually in symmetrical detached figures, seeming to be some favourite arrangement of the figures of their primary molecules; and those which, like ice, only under rare circumstances give clue to the forms of their true crystals, but habitually show themselves in accumulated mass, or complex and capricious involution. Of course the difference may be a question only of time; and the sea, cooled slowly enough, might build bergs of hexago-

nal ice-prisms as tall as Cleopatra's needle, and as broad as the tower of Windsor; but the time and temperature required, by any given mineral, for its successful constructions of form, are of course to be noted among the conditions of its history, and stated in the account of its qualities.

14. Neither, hitherto, has any sufficient distinction been made between properly crystalline and properly cleavage planes.* The first great laws of crystalline form are given by Miller as equally affecting both; but the conditions of substance which have only so much crystalline quality as to break in directions fixed at given angles, are manifestly to be distinguished decisively from those which imply an effort in the substance to collect itself into a form terminated at symmetrical distances from a given centre. The distinction is practically asserted by the mineral itself, since it is seldom that any substance has a cleavage parallel to more than one or two of its planes: and it is forced farther on our notice by the ragged lustres of true cleavage planes like those of mica, opposed to the serene bloom of the crystalline surfaces formed by the edges of the folia.

15. Yet farther. The nature of cleavage planes in definitely crystalline minerals connects itself by imperceptible gradations with that of the surfaces produced by mechanical separation in their masses consolidating

---

* See vol. i., chap. xiv., §§ 20—22.

from fusion or solution. It is now thirty years, and more, since the question whether the forms of the gneissitic buttresses of Mont Blanc were owing to cleavage or stratification, became matter of debate between leading members of the Geological Society; and it remains to this day an undetermined one! In succeeding numbers of 'Deucalion,' I shall reproduce, according to my promise in the introduction, the chapters of 'Modern Painters' which first put this question into clear form; the drawings which had been previously given by de Saussure and other geologists having never been accurate enough to explain the niceties of rock structure to their readers, although, to their own eyes on this spot, the conditions of form had been perfectly clear. I see nothing to alter either in the text of these chapters, written during the years 1845 to 1850, or in the plates and diagrams by which they were illustrated; and hitherto, the course of geological discovery has given me, I regret to say, nothing to add to them: but the methods of microscopic research originated by Mr. Sorby, cannot but issue, in the hands of the next de Saussure, in some trustworthy interpretation of the great phenomena of Alpine form.

16. I have just enough space left in this chapter to give some illustrations of the modes of crystalline increment which are not properly subjects of mathematical definition; but are variable, as in the case of the formations of ice above described, by accidents of situation, and by the modes and quantities of material supply.

17. More than a third of all known minerals crystallize in forms developed from original molecules which can be arranged in cubes and octahedrons; and it is the peculiarity of these minerals that whatever the size of their crystals, so far as they are perfect, they are of equal diameter in every direction; they may be square blocks or round balls, but do not become pillars or cylinders. A diamond, from which the crystalline figure familiar on our playing cards has taken its popular name, be it large or small, is still a diamond, in figure as well as in substance, and neither divides into a star, nor lengthens into a needle.

18. But the remaining two-thirds of mineral bodies resolve themselves into groups, which, under many distinctive conditions, have this in common,—that they consist essentially of *pillars* terminating in pyramids at both ends. A diamond of ordinary octahedric type may be roughly conceived as composed of two pyramids set base to base; and nearly all minerals belonging to other systems than the cubic, as composed of two pyramids with a tower between them. The pyramids may be four-sided, six-sided, eight-sided; the tower may be tall, or short, or, though rarely, altogether absent, leaving the crystal a diamond of its own sort; nevertheless, the primal separation of the double pyramid from the true tower with pyramid at both ends, will hold good for all practice, and to all sound intelligence.

19. Now, so long as it is the law for a mineral, that

however large it may be, its form shall be the same, we have only crystallographic questions respecting the modes of its increase. But when it has the choice whether it will be tall or short, stout or slender, and also whether it will grow at one end or the other, a number of very curious conditions present themselves, unconnected with crystallography proper, but bearing much on the formation and aspect of rocks.

20. Let *a*, fig. 1, plate X., be the section of a crystal formed by a square tower one-third higher than it is broad, and having a pyramid at each end half as high as it is broad. Such a form is the simplest general type of average crystalline dimension, not cubic, that we can take to start with.

Now if, as at *b*, we suppose the crystal to be enlarged by the addition of equal thickness or depth of material on all its surfaces,—in the figure its own thickness is added to each side,—as the process goes on, the crystal will gradually lose its elongated shape, and approximate more and more to that of a regular hexagon. If it is to retain its primary shape, the additions to its substance must be made on the diagonal lines dotted across the angles, as at *c*, and be always more at the ends than at the flanks. But it may chance to determine the additions wholly otherwise, and to enlarge, as at *d*, on the flanks instead of the points; or, as at *e*, losing all relation to the original form, prolong itself at the extremities, giving little, or perhaps nothing, to its sides. Or, lastly, it may

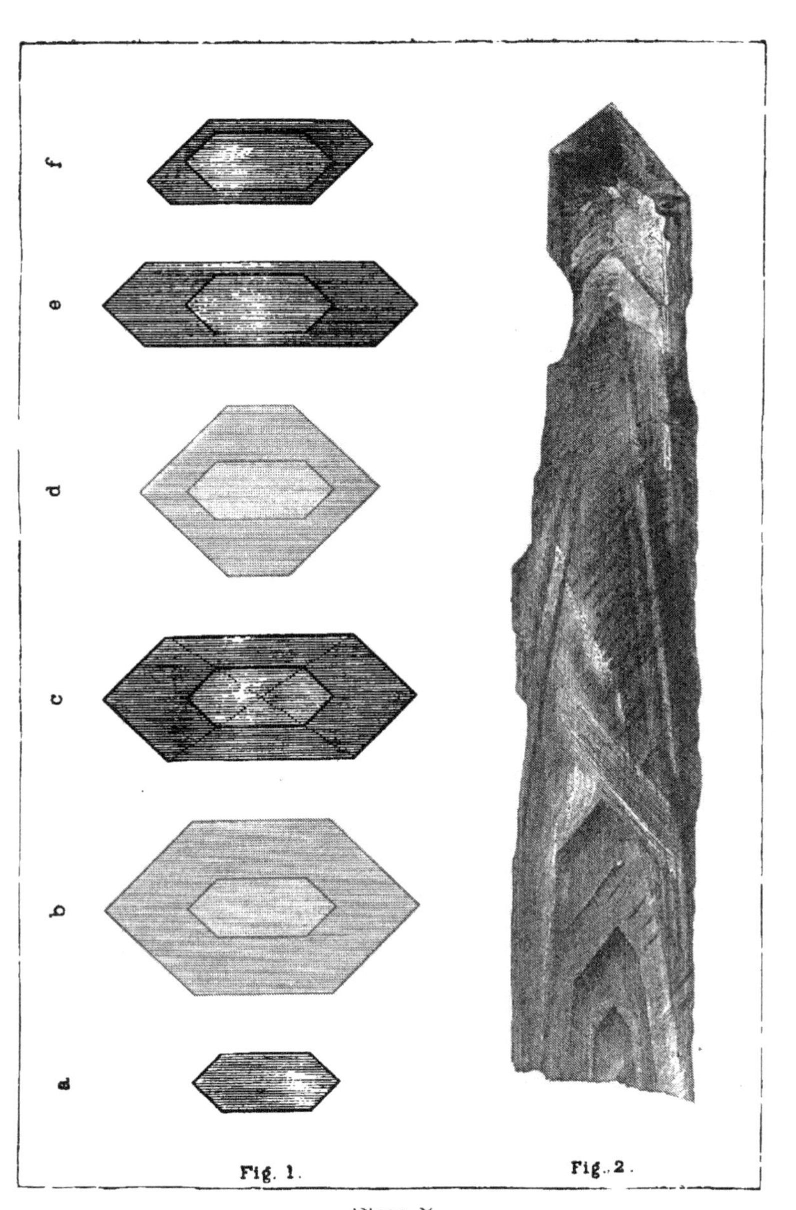


Plate X.

Modes of Crystalline Increment.

alter the axis of growth altogether, and build obliquely, as at *f*, on one or more planes in opposite directions.

21. All the effective structure and aspect of crystalline substances depend on these caprices of their aggregation. The crystal of amethyst of which a longitudinal section is given in plate X., fig. 2, is more visibly, (by help of its amethyst staining,) but not more frequently or curiously, modified by accident than any common prism of rough quartz will be usually found on close examination; but in this example, the various humors, advances, and pauses of the stone are all traced for us by its varying blush; and it is seen to have raised itself in successive layers above the original pyramid—always thin at the sides, and oblique at the summit, and apparently endeavouring to educate the rectilinear impulses of its being into compliance with a beautiful imaginary curve.

22. Of prisms more successful in this effort, and constructed finally with smoothly curved sides, as symmetrical in their entasis as a Greek pillar, it is easy to find examples in opaque quartz—(not in transparent*)—but no quartz crystal ever *bends* the vertical axis as it grows, if the prismatic structure is complete; while yet in the imperfect and fibrous state above spoken of, § 5, and mixed with clay in the flammeate forms of jasper, undulation becomes a law of its being!

23. These habits, faculties, and disabilities of common

* Smoky quartz, or even Cairngorm, will sometimes curve the sides parallel to the axis, but (I think) pure white quartz never.

quartz are of peculiar interest when compared with the totally different nature and disposition of ice, though belonging to the same crystalline system. The rigidly and limitedly mathematical mind of Cloizeaux passes without notice the mystery, and the marvel, implied in his own brief statement of its elementary form "Prisme hexagonal *regulier*." Why 'regular'? All crystals belonging to the hexagonal system are necessarily regular, in the equality of their angles. But ice is regular also in *dimensions*. A prism of quartz or calcite may be of the form *a* on the section, Fig. 6,* or of the form *b*; but ice

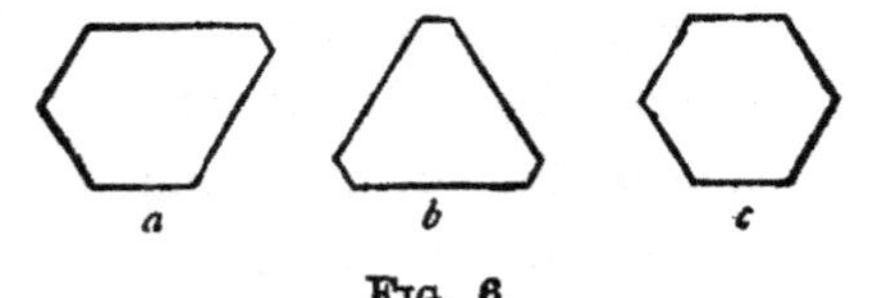


FIG. 6.

is always true—like *c*, as a bee's cell—'prisme regulier.'

So again, Cloizeaux tells us that ice habitually is formed in 'tables hexagonales *minces*.' But why thin?—and *how* thin? What proportion of surface to edge was in his mind as he wrote, undefined? The square plates of uranite, the hexagonal folia of mica, are 'minces' in a quite different sense. They can be seen separately, or in masses which are distinctly separable. But the "prisme

* I think it best to number my woodcuts consecutively through the whole work, as the plates also; but fig. 5 is a long way back, p. 166, vol. i. Some further notes on it will be found in the next chapter.

hexagonale mince, regulier" of ice cannot be split into thinner plates—cannot be built into longer prisms; but, as we have seen, when it builds, is fantastic in direction, sudden in force, endlessly complex in form.

24. Here, for instance, fig. 7, is the outline of one of the spiculæ of incipient surface ice, formed by sharp frost on calm water already cooled to the freezing point. I have seen literally clouds of surface ice woven of these barbed arrows, shot,—or breathed, across half a mile of lake in ten minutes. And every barb of them *itself* a miracle of structure, complex as an Alpine peak.

These spiculæ float with their barbs downwards, like

FIG. 7.

keels, and form guiding ribs above like those of leaves, between which the entire surface of the water becomes laminated; but, as it does so, the spiculæ get pushed up into little mountain ridges, always steeper on one side than the other—barbed on the steep side, laminated on the other—and radiating more or less trigonally from little central cones, which are raised above the water-surface with hollow spaces underneath.

And it is all done with 'prismes hexagonales reguliers'!

25. Done,—and sufficiently explained, in Professor Tyndall's imagination, by the poetical conception of 'six

poles' for every hexagon of ice.* Perhaps!—if one knew first what a pole was, itself—and how many, attractive, or repulsive, to the east and to the west, as well as to the north and the south—one might institute in imaginative science—at one's pleasure;—thus also allowing a rose five poles for its five petals, and a wallflower four for its four, and a lily three, and a hawkweed thirteen. In the meantime, we will return to the safer guidance of primal mythology.

26. The opposite plate (XI.) has been both drawn and engraved, with very happy success, from a small Greek coin, a drachma of Elis, by my good publisher's son, Hugh Allen. It is the best example I know of the Greek type of lightning, grasped or gathered in the hand of Zeus. In ordinary coins or gems, it is composed merely of three flames or forked rays, alike at both extremities. But in this Eleian thunderbolt, when the letters F.A. (the old form of beginning the name of the Eleian nation with the digamma) are placed upright, the higher extremity of the thunderbolt is seen to be twisted, in sign of the whirlwind of electric storm, while its lower extremity divides into three symmetrical lobes, like those of a flower, with spiral tendrils from the lateral points: as constantly the honeysuckle ornament on vases, and the other double groups of volute completed in the Ionic

---

* 'Forms of Water,' in the chapter on snow. The discovery is announced, with much self-applause, as an important step in science.

Plate XI.

The Olympian Lightning.

REESE LIBRARY
OF THE
UNIVERSITY
CALIFORNIA.

capital, and passing through minor forms into the earliest recognizable types of the fleur-de-lys.

27. The intention of the twisted rays to express the action of storm is not questionable—"tres imbris torti radios, et alitis austri." But there can also be little doubt that the tranquillities of line in the lower divisions of the symbol are intended to express the vital and formative power of electricity in its terrestrial currents. If my readers will refer to the chapter in 'Proserpina' on the roots of plants, they will find reasons suggested for concluding that the root is not merely a channel of material nourishment to the plant, but has a vital influence by mere contact with the earth, which the Greek probably thought of as depending on the conveyance of terrestrial electricity. We know, to this day, little more of the great functions of this distributed fire than he: nor how much, while we subdue or pervert it to our vulgar uses, we are in every beat of the heart and glance of the eye, dependent, with the herb of the field and the crystal of the hills, on the aid of its everlasting force. If less than this was implied by the Olympian art of olden time, we have at least, since, learned enough to read, for ourselves his symbol, into the higher faith, that, in the hand of the Father of heaven, the lightning is not for destruction only; but glows, with a deeper strength than the sun's heat or the stars' light, through all the forms of matter, to purify them, to direct, and to save.

CPSIA information can be obtained
at www.ICGtesting.com
Printed in the USA
LVOW13s0359181017
552845LV00014B/369/P

9 781376 321661